The Water Cycle

Theresa Greenaway

HODDER
Wayland
an imprint of Hodder
Children's Books

CYCLES IN NATURE

Other titles in this series:

Food Chains

Plant Life

All Hodder Wayland books encourage children to read and help them improve their literacy.

 The contents page, page numbers, headings and index help locate specific pieces of information.

 The glossary reinforces alphabetic knowledge and extends vocabulary.

✓ The further information section suggests other books and websites dealing with the same subject.

Cover: A drop of water splashing in a puddle (*main image*); (*clockwise from top right*) clouds; a diagram of the water cycle; a tiger drinking; drops of water on a twig.
Title page: A drop of water splashing in a puddle.
Contents page: A girl in Mali enjoying a drink of water from an outdoor pump.

Series editor: Nicola Wright
Book editor: Alison Cooper
Series and cover design: Sterling Associates
Book design: Jean Wheeler

First published in 2000 by Hodder Wayland, an imprint of Hodder Children's Books
This edition published in 2001

© Hodder Wayland 2000

Typeset by Jean Wheeler
Printed and bound in Grafiasa, Porto, Portugal

British Library Cataloguing in Publication Data
Greenaway, Theresa
The Water Cycle - (Cycles in nature)
1. Hydrologic cycle - Juvenile literature
2. Water - Juvenile literature
3. Plant-water relationships - Juvenile literature
4. Animal-water relationships - Juvenile literature
I. Title
551.4'8

ISBN 0 7502 3471 7

Picture acknowledgements
The publishers would like to thank the following for allowing their images to be used in this book: Bryan & Cherry Alexander 7; Axiom 4; Biofotos 13 (*top*)/Soames & Summerhays; Bruce Coleman *cover (top left) cover (bottom left)* /Oasis Inc.; Eye Ubiquitous 9/R.D. Battersby; Robert Harding *cover (top right)*, 10,14; Oxford Scientific Films 5/Richard Packwood, 13 (*bottom*)/ Ben Osborne, 16 (*top*)/ Mike Hill, 27/Richard Packwood; Tony Stone Images 6/Tom Till, 8, 12/Paul Harris, 19/Philip Condit II; 25/Bruce Forster, 26/David Woodfall, 28/Colin Raw, 29/Chris Simpson; Panos contents page & 24/Giacomo Pirozzi; Pictor *cover (main image)* & *title page*; Zefa/Stock Market 16 (bottom). Artwork on pages 15, 17 (*inset only*) and 20–21 by Peter Bull; all other artwork from Wayland Picture Library.

Contents

Water all Around

Look around you. Where can you see water? At home or school, water gushes from a tap whenever you need it. On a hot summer's day, you might not be able to see any water outside, but it is still there, inside plants and animals. There is even water inside you! When it rains, water seems to be everywhere. It drips from leaves and umbrellas and makes puddles on the road.

◄ A torrential downpour during the monsoon rains in Hô Chi Minh, Vietnam.

Water on the move

The water around us is always on the move. It moves through our bodies when we have a drink. It travels from the roots of a plant to the topmost leaf.

Rain falls into streams that join to make a river. Eventually, rivers flow into the sea. Water evaporates from the sea and rises high into the air as water vapour. This movement of water is called the water cycle. You can find out more about it on pages 10–11.

▲ Water tumbles down a waterfall in Iceland into a mountain stream.

How much of your weight is water?

Did you know that 70 per cent of your body weight is water? Weigh yourself on bathroom scales to find out how heavy you are. Divide that figure by ten and then multiply by seven to find out what 70 per cent of your weight is. You can work out how much water you and all your classmates contain.

Changing States

Water is liquid between 0 °C and 100 °C. At 0 °C it freezes into a solid called ice. Above 100 °C, it changes into a gas called water vapour.

▼ Icebergs are massive, floating chunks of ice that are found in the seas around the North and South Poles.

Floating ice

When liquids cool, the molecules from which they are made up squeeze together more tightly. In other words, the liquids become more dense. They freeze, or become solid, from the bottom up.

Water is different from all other liquids. When it cools, the molecules squeeze together until the water temperature falls to 4 °C, but then they spread out again. When water freezes, it is much less dense than liquid water. This means that solid water, or ice, is lighter than liquid water, and so it floats.

▲ Harp seals like these swim under the ice to catch fish.

When a layer of ice forms on a lake or stream, some water below it usually remains liquid. Fish and other water creatures can stay alive in the water, until the warm sunshine melts the surface ice. If water froze from the bottom up, these animals would be trapped in the ice, and they would die.

Floating and sinking solids

Pour water into one half of an ice-cube tray and cooking oil into the other half. Put the tray in the freezer and leave it until the water and oil are frozen solid. Half-fill one beaker with cold water, and another with cooking oil. Drop an ice cube into the beaker of water and an oil cube into the beaker of oil. Which one sinks and which one floats?

Evaporation

As liquid water becomes warmer, it escapes into the air as a gas called water vapour. This is called evaporation. On a hot day, water evaporates quickly. Even when it is cold, some evaporation still takes place, so there is always some water vapour in the air.

The water in puddles and small ponds evaporates until there is none left. Water evaporates from the sea all the time, too, but as it is always being replaced by rain and river water, the sea never dries up.

▲ Water is evaporating rapidly from this waterhole in Namibia.

Investigating evaporation

Wet clothes dry when the water evaporates from them. In what kind of weather do you think water evaporates fastest? To find out, wet a clean handkerchief and squeeze it until no water drips out. Peg it on a line outside. Time how long it takes to dry when the weather is warm, cool, damp or windy.

When water is heated to 100 °C, it boils. Large bubbles appear and the liquid water turns to water vapour. We call this hot water vapour steam. You can see steam when you boil a kettle. In some parts of the world you can see steam gushing out of the ground. This is called a geyser. Geysers form where water is heated by hot rocks deep underground.

Condensation

Steam soon cools down as it meets the air. It turns back into tiny droplets of liquid water. This is called condensation. It is the exact opposite of evaporation.

A geyser in Yellowstone National Park, USA, shoots a jet of steam high into the air. ▶

The Water Cycle

The movement of water in the environment is called the water cycle. Warmth from the sun causes water to evaporate from seas, lakes and rivers, and from the leaves of plants. The warm water vapour rises into the air. As it rises, it cools and condenses to form clouds.

Clouds that form over the oceans rise when they are blown over hills and mountains on land. They become even cooler. The tiny droplets of water join up and eventually fall from the clouds as rain, snow or hail. Rain fills up streams, lakes, rivers and reservoirs. Some of it soaks down through the soil and collects in water-holding rocks called aquifers. People take water from lakes, rivers, reservoirs and aquifers. We use this water in our homes, farms and factories.

▲ Clouds form when water vapour condenses high in the sky.

Rivers flow into the sea. The water that evaporated from the sea has been returned to it, completing the water cycle.

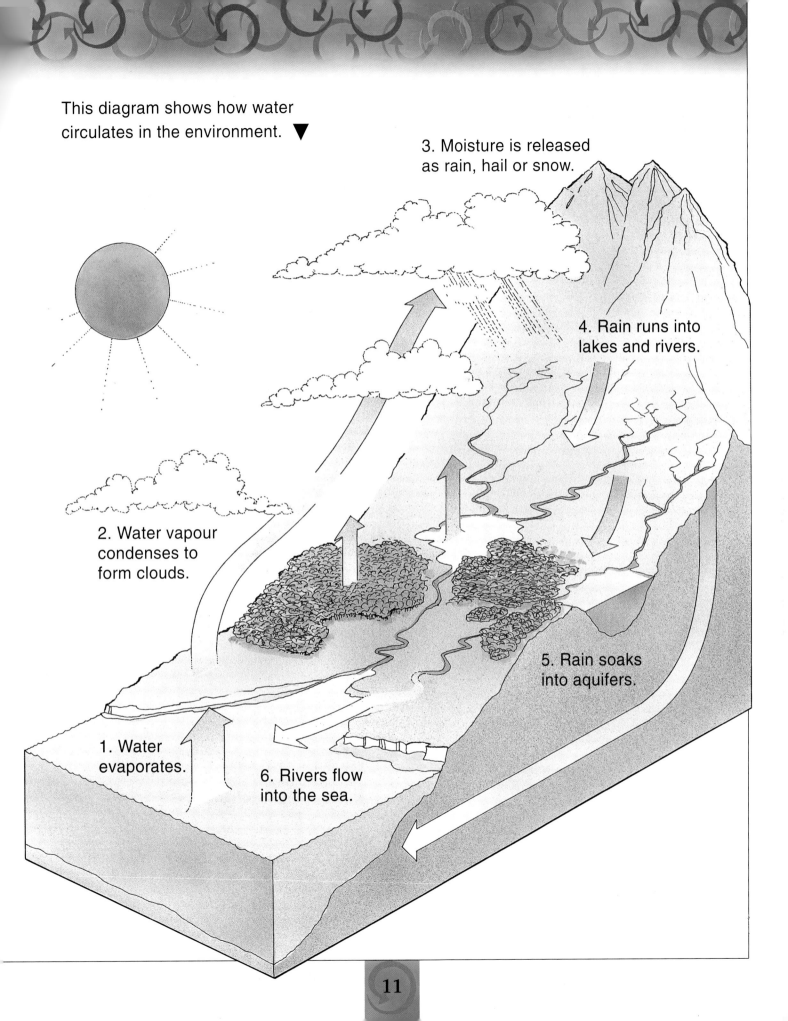

Water around the world

As the warm sun shines down on trees and smaller plants, water evaporates from their leaves. We do not notice the water vapour lost from plants in our garden. But huge amounts of water vapour rise from a tropical rain forest every day. The water vapour condenses to form clouds and there are heavy downpours of rain late in the afternoons.

▼ Water vapour rising from this rain forest in Borneo has condensed to form clouds.

Deserts are places where there is less than 25 cm of rain a year. Often, deserts are found behind high mountain ranges. When warm, moist air rolls in from the sea, it rises and cools as it meets the mountains. All the water vapour condenses and falls as rain. By the time the air has passed over the mountains, it has lost all its moisture.

◀ Cacti are able to survive in the hot, dry desert of California, USA.

The North and South Poles are the two points on the Earth that are furthest away from the Sun. These areas are bitterly cold all year round. In the north, the Arctic Sea is permanently frozen. The South Pole is surrounded by a huge frozen continent – Antarctica. Even in summer, this land is covered by an ice sheet up to 3 km thick.

The icy seas around Antarctica never thaw completely. ▶

Water for Life

All living things – plants, animals and people – need water. We cannot feed, breathe, grow or reproduce without it.

Animals and people

Land animals need to drink water. Once it has been swallowed, some of the water stays in the digestive system. The rest passes through the wall of the stomach and intestines and travels to all the other parts of the body.

Water leaves an animal as either a vapour or a liquid. Water vapour is lost every time we breathe out. Water also evaporates from our bodies in the form of sweat. This helps us to keep cool on a hot day. People can sweat all over their bodies. A cat only sweats from its pads and nose.

◀ A dog opens its mouth and pants so that water can evaporate from its large tongue.

Seeing your breath

You can see that the air you breathe out contains water vapour if you breathe on to a cold mirror. The water vapour condenses on its surface.

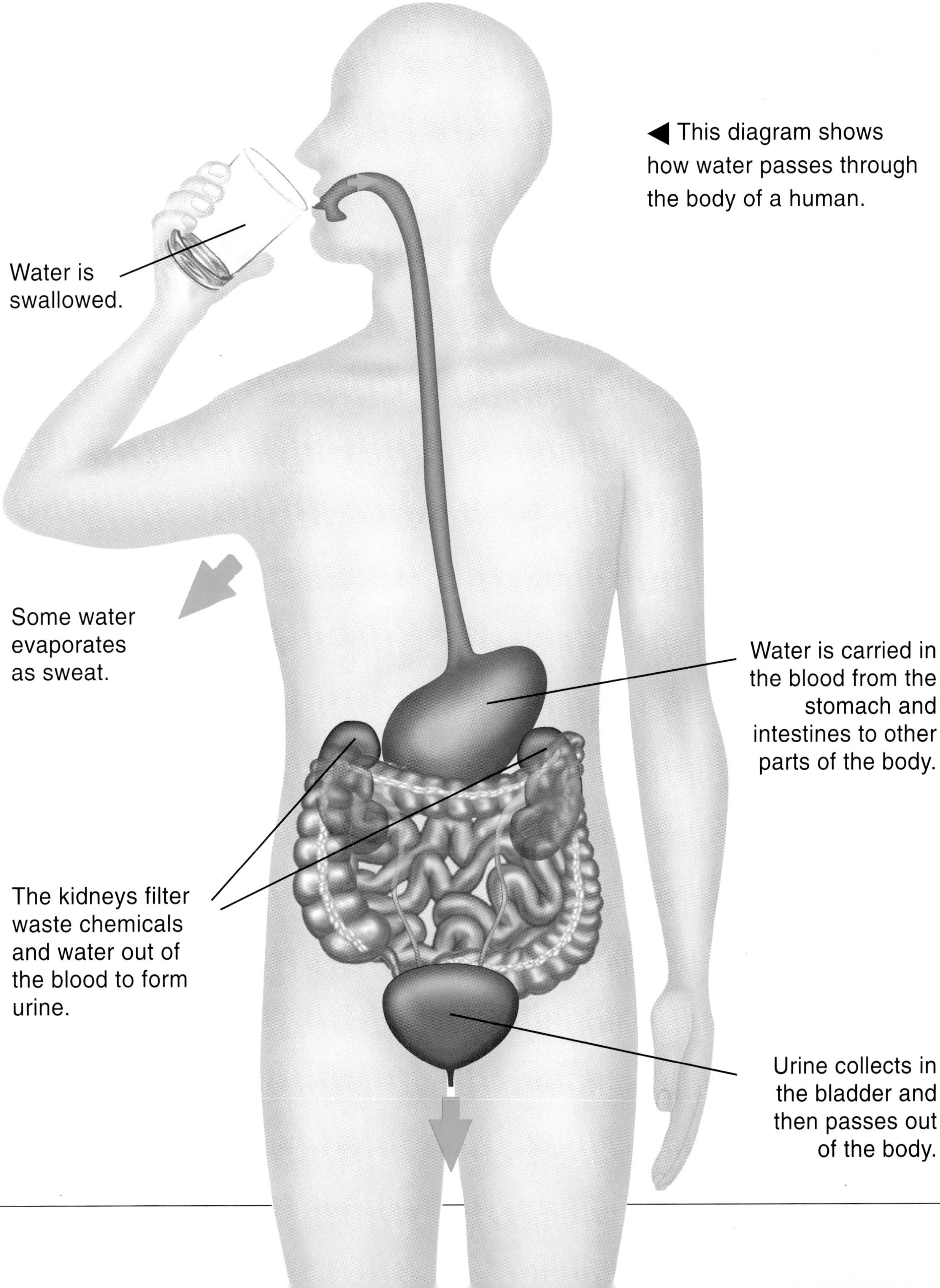

Water is
swallowed.

◀ This diagram shows
how water passes through
the body of a human.

Some water
evaporates
as sweat.

Water is carried in
the blood from the
stomach and
intestines to other
parts of the body.

The kidneys filter
waste chemicals
and water out of
the blood to form
urine.

Urine collects in
the bladder and
then passes out
of the body.

Surviving in dry places

Finding enough water to drink is hard for people and animals that live in dry places such as deserts. Months, or even years, can go by with no rain.

Desert animals have developed ways of making the best use of the scarce water supply. Most small desert animals stay underground, where it is cooler, during the day. This reduces the amount of water that evaporates from their bodies. They come out at night and drink droplets of dew.

People and animals depend on wells and oases for their water. Water springs up naturally at an oasis but may have to be pumped up from a deep well.

▲ Jerboas do not sweat at all. Their urine contains very little water.

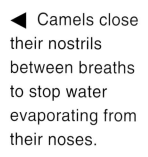

◄ Camels close their nostrils between breaths to stop water evaporating from their noses.

Plants

Plants absorb water through their roots. It is carried up through very thin tubes into the leaves, shoots, buds and flowers. Just like animals, plants need water to grow and reproduce. Water is also needed for photosynthesis, the process in which plants make sugars from sunlight and carbon dioxide.

▼ This diagram shows how water passes through a plant.

A magnified ▶ view of the inside of a leaf.

Water passes out of the leaf vein into the surrounding cells.

Water evaporates through tiny holes in the leaf.

Water evaporates from the leaves

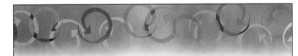

How quickly does water travel up a stem?

Put some drops of red or blue food colouring into water in a plastic beaker. Stand a few sticks of celery in the beaker. How long does it take for the coloured water to travel all the way up the celery stems?

Thin tubes carry water up the stem.

Roots absorb water from the soil.

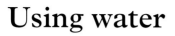

Using water

Water vapour is lost through tiny holes in a plant's leaves. As the water evaporates, more water is drawn up the plant from its roots. It can be pulled right to the top of the tallest tree. But if water vapour escapes from the leaves faster than water can be drawn up from the soil, then the leaves become floppy, or 'wilt'.

▼ A healthy plant (left) and a wilted plant (right).

Do plants lose water vapour?

You will need two leafy, green plants in flower pots. Make sure that the soil in both pots is moist. Tie a polythene bag around the base of one of the plants, so that the flower pot stays outside the bag. Leave both plants for a day. Condensation will appear on the polythene bag, but you cannot see it on the other plant.

Although plants can make their own sugars, they need many other mineral nutrients in order to grow. Minerals in the soil, such as nitrates and phosphates, dissolve in water. They are taken in when the roots absorb water.

Desert plants

Just like people and animals that live in dry places, desert plants have developed ways to conserve water. Cacti grow in the deserts of Mexico and the southern USA. Their leaves are no more than white spines, so that there is very little surface area from which water can evaporate. At night, moisture from the air condenses on the spines and trickles off as droplets of dew that soak into the ground around the cactus roots.

A cactus stores rain ▶ water in its fleshy stems.

Recycling Water

The water in lakes and rivers is called 'freshwater' but that does not mean it is safe to drink. It might contain organisms that can cause disease, or poisonous chemicals. Disease-causing organisms can be killed by boiling water. But to make sure there is enough safe water for millions of people to drink, water is pumped out of rivers and reservoirs and passed through a water purification system.

Water purification system

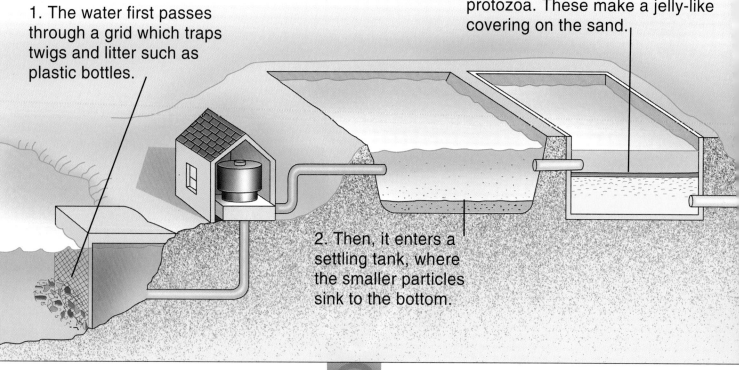

3. The next stage is to pass the water through a filter. This is made of sand covered with a layer of simple organisms called protozoa. These make a jelly-like covering on the sand.

1. The water first passes through a grid which traps twigs and litter such as plastic bottles.

2. Then, it enters a settling tank, where the smaller particles sink to the bottom.

Making the best water filter

Fold a circle of filter paper to fit inside a plastic funnel and put the funnel over a beaker. Tip some stony soil into a beaker of water and slowly pour it through the filter paper. How much of the soil has the paper removed? What happens if you put a spoonful of sand into the filter paper before you pour the dirty water through? What happens if you use a spoonful of gravel? Try using the sand, gravel and filter paper together. Have you improved your filter?

5. The purified water is stored in an enclosed reservoir or water tower. Houses, offices, schools and hospitals are all supplied with this clean, safe water.

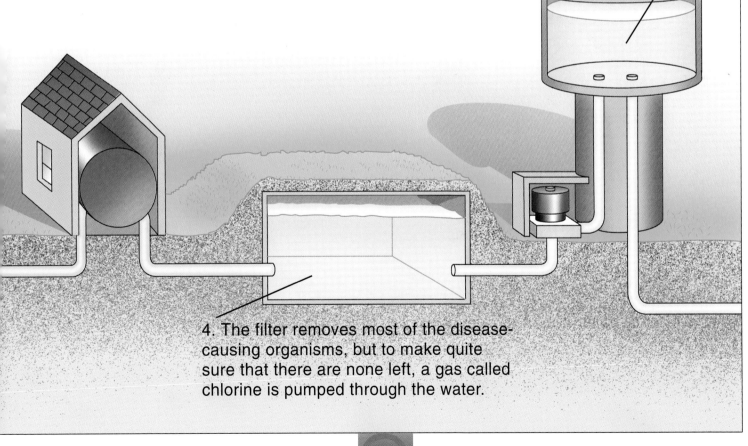

4. The filter removes most of the disease-causing organisms, but to make quite sure that there are none left, a gas called chlorine is pumped through the water.

Waste water

The wastes from wild animals – urine and faeces (droppings) – pass naturally back into the soil. Waste water produced by people contains household waste, water from industry, and sewage – the urine and faeces that we produce. If this waste water is returned directly to rivers or lakes, it causes serious pollution. Detergents and industrial chemicals damage wildlife. If we were to drink this water, the disease-causing organisms in sewage would make us ill.

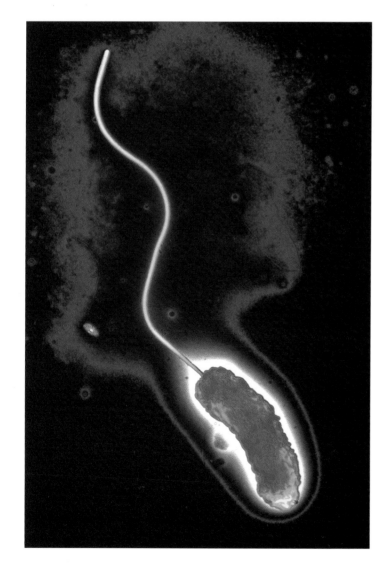

Sewage treatment

Waste water passes along underground pipes to sewage treatment plants. In cities and built-up areas all round the world, the amount of waste water and sewage produced every day is enormous. This waste is treated by the activated sludge process.

◀ This is a cholera bacterium, seen through a microscope. Cholera is a disease-causing organism found in water that has been polluted by faeces.

When sewage and waste water arrive at the sewage works, they pass through a grid which removes large bits of rubbish. The sewage then slowly flows through a series of tanks. In these tanks, tiny organisms feed on the sewage and on harmful disease-causing organisms. At the end of the process, the water is clean enough to be returned to a river or to the sea.

▲ Bacteria in these filter beds break down sewage to make it harmless.

Human Interference

Water is so much part of our daily lives that we sometimes take it for granted. We turn on a tap at home and out gushes clean, safe water. People in some developing countries, who may have to walk long distances to fill containers with clean water from wells, understand that water is precious. The way in which we use water can have a serious effect on our environment and on the water supplies that may be available for us in the future.

Using too much water

In the USA, every person uses an average of about 600 litres of water every day. Yet people in less developed countries use only about 90 litres a day. Each day, factories and power stations in developed countries use millions of kilolitres of water. Most of this water is taken from rivers or reservoirs. In many places, so much water is being used that rivers are in danger of drying up, especially in summer.

◀ A girl in Mali enjoys a cool drink of clean water from an outdoor pump.

Water is likely to become even more scarce as the human population grows. We are using it faster than it can naturally recycle. Unless we can be more careful with water, there will not be enough to go round.

▲ These sprinkler systems are spraying vast amounts of water on to farm crops.

How much can you save?

Measure how much water you use for different tasks, such as cleaning your teeth, washing hands or painting a picture. Then try to work out how you might use less. For example, you can save water by turning the tap off instead of leaving it running while you clean your teeth. But remember it is still important to make sure your hands are clean before preparing or eating food!

Pollution

Many chemicals that are made in factories and used on farms are very poisonous. If these are deliberately or accidentally released into water, plants and animals may be killed.

Oil that spills from damaged tankers out at sea floats on the surface. Seabirds, whales and seals are all injured or killed if they become covered in the oil or swallow it.

Finding out more

How many different pollutants can you think of? Search through books, newspapers, CD-ROM encyclopaedias or the Internet to find information about water pollutants and the damage they can cause. You could use your information for a project about water pollution. But remember – keep away from polluted water, it is dangerous for you as well as for wild plants and animals.

▼ Water polluted with chemicals pours from a factory waste pipe into a stream.

Many industries produce waste gases called sulphur dioxide and nitrogen oxides. Power stations that produce electricity from coal, for example, release large amounts of these gases into the atmosphere. They dissolve in falling rain to make dilute acids. This acid rain kills trees.

▼ These trees in Poland have been killed by acid rain.

Global warming

The atmosphere around the Earth acts like a greenhouse. It traps some of the Sun's heat, making our planet warm enough for living things to survive. But many scientists believe that the Earth is gradually becoming hotter and hotter. This is called global warming.

The increase in temperature is being caused by an increase in the amounts of gases such as carbon dioxide and methane in the atmosphere. These gases have become known as 'greenhouse gases' because they are trapping more of the Sun's heat. Carbon dioxide is released when fossil fuels – coal, oil and natural gas – are burnt. Coal, oil and gas are burnt in power stations to make electricity. Cars and aeroplanes also use millions of litres of fuel every day. Other industries also release greenhouse gases.

Global warming could have dramatic effects on the world's climate and water supplies. Some areas will get drier. Others may become wetter and warmer. Some plants may no longer be able to grow in certain areas, so farmers will have to change the crops they grow.

Seawater will expand as it warms and the polar ice-caps may melt, so sea-levels could rise. Low-lying coastal areas may become permanently flooded. No one knows exactly what effects global warming will have on wildlife.

◀ Power stations release large amounts of greenhouse gases into the atmosphere.

Coral reefs like this one are dying because tropical seas are becoming warmer. ▶

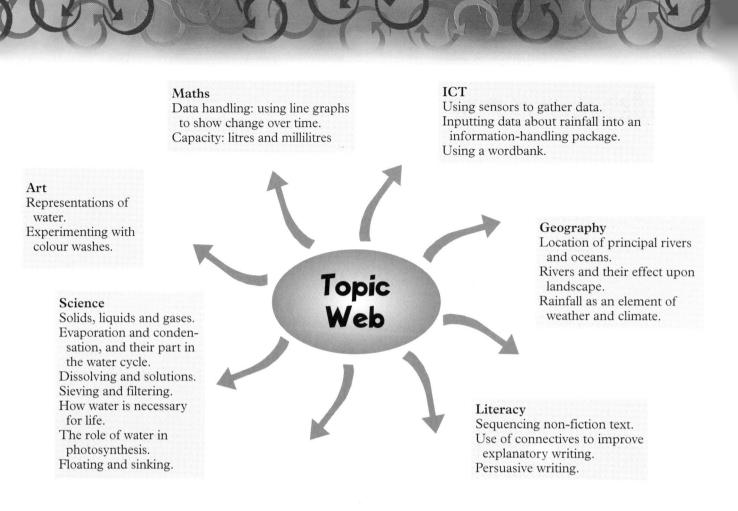

Maths
Data handling: using line graphs
to show change over time.
Capacity: litres and millilitres

ICT
Using sensors to gather data.
Inputting data about rainfall into an
information-handling package.
Using a wordbank.

Art
Representations of
water.
Experimenting with
colour washes.

Geography
Location of principal rivers
and oceans.
Rivers and their effect upon
landscape.
Rainfall as an element of
weather and climate.

Science
Solids, liquids and gases.
Evaporation and conden-
sation, and their part in
the water cycle.
Dissolving and solutions.
Sieving and filtering.
How water is necessary
for life.
The role of water in
photosynthesis.
Floating and sinking.

Topic Web

Literacy
Sequencing non-fiction text.
Use of connectives to improve
explanatory writing.
Persuasive writing.

More Activities

Art
• Look at how different artists have represented
water in paintings and drawings. You might like
to look at paintings by Turner; paintings by
Impressionists such as Signac, Seurat, Whistler
or Monet; colour woodprints by the Japanese
artist Hiroshige.

Literacy
• Make a leaflet to persuade other people to be
careful with water. Compare your leaflets with
real-life leaflets about water conservation.

Science
• To help you understand the difference between
mass and density, and their relevance to floating
and sinking, you can investigate whether a selection
of objects float or sink. Weigh each object and
predict whether it will float or sink. You might
predict that all the heavier objects will sink, but you
will discover that in fact some large, heavy objects
float, because they are less dense than water. You
can investigate this idea further with two identical
bags, filled with the same mass of sand and flour –
one denser than water, the other less dense.

Maths
• Use a measuring beaker to collect rainfall outside
your school. Measure the amount of rainfall each
day for a month and record your results on a line
graph. Try measuring the rainfall at your home
too. Can you record both sets of results on the
same graph?

• How much water is there in a litre? Measure out a
litre of water to give you an idea of what that
amount looks like. Investigate packaging that shows
measurements in litres and millilitres. For example,
how many juice cartons would you need to hold the
liquid from a big bottle of cola?

ICT
• Use sensors in a cup of ice to measure changes in
temperature as the ice melts. Record your results
on a line graph.

• Input data about annual rainfall in different
countries on to a data base. Print out the data in
different formats – which are clearer?

Glossary

acid A chemical that can burn or 'eat away' other substances.

aquifers Underground rocks in which water collects naturally.

bacterium A living thing that can only be seen through a microscope. The plural is bacteria.

climate The type of weather in a particular area over many years.

condenses Changes from a gas to a liquid.

conserve To prevent something from being lost.

detergents Chemicals used for cleaning.

developing countries Poorer countries which do not have many industries.

digestive system The parts of the body that work together to break down food so that it can be used for healthy growth.

evaporates Changes from a liquid to a gas.

filter Something that allows some materials to pass through but not others.

intestines Tubes inside the body that are part of the digestive system.

minerals Chemical substances.

molecules Very tiny parts from which all objects are made up.

nutrients Substances that all living things need for healthy growth.

organisms All living things.

reservoirs Natural or artificial lakes used for storing water.

water vapour The gas that forms when water evaporates.

Further Information

BOOKS

The Earth Strikes Back: Water by Pamela Grant and Arthur Haswell (Belitha Press, 1999)

The Way it Works: Water (Heinemann, 1998)

Threads: Water (A & C Black, 1998)

WEBSITES

www.sas.org.uk. Information from campaigning environmental group, Surfers Against Sewage, on marine pollution and toxic waste discharges.

www.ce.vt.edu/enviro2/wtprimer.html. Explanation of processes that go on at your local waterworks from Virginia Tech's Water Treatment Primer.

www.hydro.co.uk. Scottish Hydro-Electric's information on water as a source of power.

www.fhc.co.uk. Information on water as a source of power in Wales from First Hydro.

www.lcra.org. The Lower Colorado River Authority has pictures of dams and a useful glossary of water terms.

Index

Page numbers printed in **bold** mean that there is information about this topic in a photograph, diagram or caption.